Blue Does Not Exist

By
John Cantellow

Also by John Cantellow

Beyond Forgiveness to
unconditional love

Discovering Journey's End our
final destination

Pure Love a new paradigm

Journey into Pure Love

Blue does not exist

Pure Love Theology

Pure Love Ideology

Think! Earth

Life Bubble
viewed from outside

Contents

1 Introduction

If I say, "Blue does not exist", you might reasonably reply, "just look up at the sky, dummy!" However, if you were of the Himba tribe in Namibia, you would not do so, since they are unable to distinguish the color blue, though there is no physiological reason for this. They can, however, distinguish more shades of green than the wider population.

 Also, if we were living in ancient times the question would not arise since the color blue does not appear in any of the ancient texts, eg Homer's Odyssey (8th century BCE). It first appeared in Egyptian writing about 2,500 BCE. Furthermore, the inability to distinguish the color green is a common form of color blindness amongst males.

But sometimes things seem to pop into our heads totally out of the blue, so to speak. One day this was what popped into my head, "Blue does not exist." This struck me as a very strange thing to think. Because it was so odd, I couldn't just dismiss it. At about that time I had been watching a series of documentary programmes tracing the great rivers from their source to the sea, eg the Nile, the Amazon and the Ganges. So, I decided to do the same for the color blue.

The most obvious place for me to start was to look up at the sky on a cloudless, sunny day. From horizon to horizon I saw nothing but blue. The blue effect arose from the visual part of the electromagnetic spectrum that had originated from the sun being refracted by the Earth's atmosphere such that it appeared to be blue.

The photons of light then continued on their path until they hit my eye. They traveled through my eye eventually causing a stimulus to my optic nerve at the back of my eye. The optic nerve conveyed that signal all the way to the back of my brain, where vision is initially processed, and then to the ventral occipital lobe where color perception is determined.

But here's the curious thing. If we were to look closely at the photons traveling all the way from the sun and eventually causing a reaction in the ventral occipital lobe the color blue does not appear anywhere in that journey.

Photons are mediators of charge and magnetism but are not themselves blue. They are still not blue when they hit the eye, nor when traveling through the eye. Nor is the electrochemical signal traveling through the optic nerve to the back of the brain blue. And if we were to look closely at the brain, including the final destination of the ventral occipital lobe, we would still not see any blue.

2 Subjective Consciousness

And yet, unquestionably, we experience the color blue. How come? This is what's become known as, "the hard question of consciousness", and there are many researchers trying to fathom out how it is that we experience consciousness in the way that we do. To quote David Chalmers', Australian philosopher and cognitive scientist, "Why is there this inner subjective movie?" and "Why is it that all this behavior is accompanied by subjective experience?" and "Why does all this feel like something from the inside?"

For me, strange though it may seem, brussels sprouts might hold something of a clue. You see for some unfortunate people, and you might be one, brussels sprouts have a particularly bitter taste. These people have the gene variant that, when expressed, produces taste buds that are particularly sensitive to

6-*n*-propylthiouracil, the chemical responsible for the bitter taste.

Taste, just like the color blue, is a subjective conscious experience. So, the reason that photons, vibrating at a particular frequency, eventually trigger a response in our ventral occipital lobe that we experience as the color blue, may be down to our genes.

But that is still only part of the story. So far, we have established that the color blue does not physically exist. But we haven't said anything about the color blue that we actually experience. How can we experience something that doesn't actually exist?

Here it is helpful to turn to the analogy of holograms. I say analogy but it might be closer to the actual mechanism than that. First, we must distinguish between real holograms and visual effects called holograms, but which aren't. In the UK, the most widely used and common experience of pseudo holograms is as a security feature on bank notes, bank and credit cards, and drivers' licenses. These are made using two or more layers of images such that as the angle of view changes so the image layer viewed changes and the object in the image appears to move.

Then there is the theatrical pseudo hologram such that a person appears on stage but who is not physically present on stage. The same technique is used in Amusement Parks to project an image such that a ghost appears to be seated next to someone in say a haunted house or a Ghost train. This method is called Pepper's Ghost and is achieved by a bright light shining on a person or image, located beneath a stage for example, which is then projected via an angled mirror onto another surface up on stage.

Real holograms are created by quantum
interference patterns.

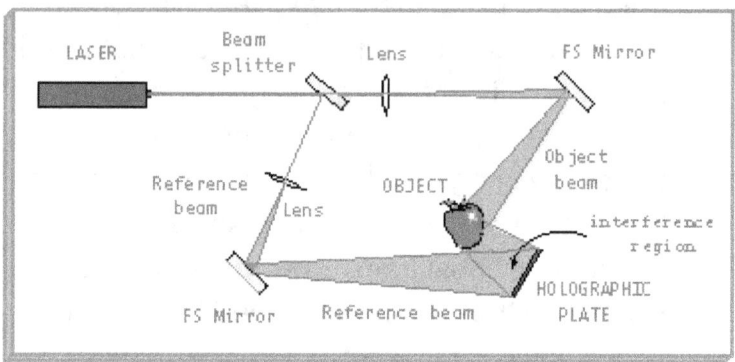

A light beam from a laser impinges on a beam
splitter that causes 50% of the light to pass
directly through and 50% of the light to be
reflected. The beam that passes directly
through is then used to illuminate the object
whose image we wish to capture.

Having been reflected from the object, it hits a
holographic plate where it meets the beam that
was reflected from the beam splitter. This
creates a quantum interference pattern on the
holographic plate. Not only does the
interference pattern contain all of the
information necessary to recreate a 3D image
of the original object but also even a small part

of the holographic plate can be used to recreate the entire image, albeit in less detail.

The final holographic image, or hologram, is created by shining the laser light onto the holographic plate. If you have seen real holograms you will know that they appear to float eerily in space. They are clearly not real physical objects. If you reached out your hand there would be nothing there to touch, and yet you can see the original object in great detail as a 3D image. In that sense it is an optical illusion.

What we experience as the color blue is an optical illusion, in much the same way that a hologram is an optical illusion. Of course, the story doesn't end with just the color blue and we shall be exploring the wider implications later.

But the question remains, why do we experience the color blue and not any other color, or indeed even a sound? Those with Chromesthesia, a type of Synesthesia, experience sounds as color. We shall return to this question in a later chapter.

Now let us consider the matter of pain. It has been long known that pain is a subjective experience and as we have already established that means it does not physically exist. The source of pain undoubtedly physically exists but the experience of pain does not. This might be why it is possible for patients to undergo surgery without anesthetic, using only hypnosis.

3 What is Information?

The color blue is information. It tells us something more about an object. We could send someone to buy a dress, but if we sent them to buy a blue dress we have provided more information about the type of dress we require. Information plays a key role in our lives. We use it all of the time. But what exactly is information?

The study of information dates back to the Ancient Greeks. They reasoned that it was events that created information. Furthermore, as Aristotle pointed out, the more surprising an event was the more information it created.

This is a key concept in Information Theory, that the quantity of information created is inversely proportional to the probability of an event occurring. So, an unlikely event with a low probability of occurring creates more information than an expected event with a high probability of occurring.

That is relatively straightforward for a single event but what about multiple, unrelated events? Obviously, the total amount of information created will be the sum of the information created for each individual event.

Fortunately, there is an easy way to add together the results of the individual calculations by using the mathematical tool of logarithms. So, the accepted definition of information in Information Theory is given by the formula

$I = \log 1/p$

Where I is the quantity of information created and p is the probability of the event occurring.

So, the quantity of information created is equal to the logarithm of the inverse probability of the event happening. For multiple, unrelated events simply add the results together to get the total information created.

This formula is attributed to Claude Shannon, whilst an engineer at Bell laboratories, and was included in his 1948 paper that forms the basis of modern Information Theory. Shannon thus established himself as the father of Information Theory and applied his knowledge to optimizing telephone channel capacity, enabling multiple telephone calls to share the same line.

If we apply Shannon's formula to tossing a coin we can see that there are only two possible outcomes, heads or tails, and that each has a probability of 0.5 of occurring. If, instead of just one coin, we have two coins the probability of both producing heads would be

$0.5 \times 0.5 = 0.25$

The quantity of information created would be

$I = \log 1/p$, ie $\log 1/0.5 = 0.3010$ for a single toss

So, for two tosses the quantity of information created would be

$0.3010 + 0.3010 = 0.6020$

And $\log 1/0.25 = 0.6020$

Information stored in computers, and more recently communicated via telephone networks and transmitted for television and radio, uses binary digital coding. Basically, information is converted into a series of 0's and 1's.

In the 1960's a physicist at IBM, Rolf Landauer, together with Charles Bennett proved that a computer must heat up as it functions. Heat is a form of physical energy. To quote Vlatko Vidral (2010, p74), "the main message from Landauer and Bennett's work is that information, rather than being an abstract notion, is entirely a physical quantity."

But, as we have already seen, the color blue does not physically exist and yet it is unquestionably information. How can this be reconciled with accepted Information Theory, which asserts that information is entirely a physical quantity?

Information Theory, from Shannon through to Landauer and Bennett, deals with physical information. That is to say information that has been encoded into a physical medium. So, it is hardly surprising that they say that information is physical.

There is no question that it is information. Let me illustrate the key difference between physical information and information that does not physically exist by two examples.

My house has a South facing bedroom with large windows and so, at times, becomes very hot. My lounge faces North, also with a large glass area, and so, at times, becomes very cold.

I have installed temperature sensors in both rooms connected to an Arduino microprocessor, programmed such that when the respective threshold temperatures are reached the microprocessor switches on a pump to transfer the hot air from the bedroom to the lounge.

It is clear that the temperature sensors are communicating information to the microprocessor and that that information takes the physical form of an electrical current.

Now consider this. Alice had told Bob that she wanted him to buy her a dress but she hadn't told him what color she wanted. Alice had the color in her subjective consciousness and so picked up a pen and drew symbols onto a piece of paper. Observers would have said that she was writing.

She then sent the piece of paper to Bob. Bob looked at the markings on the piece of paper and saw *Blue*. Observers would have said that

Bob was reading. Bob's brain decoded the markings on the paper such that his subjective consciousness understood the color blue.

In the first example the velocity of air molecules in the bedroom causes the sensor to register a temperature, which it then encodes into an electrical signal to send to the microprocessor. The microprocessor compares that with a similar encoded signal from the sensor in the lounge to determine if the pump should be switched on. Throughout we are dealing with information in physical form.

By contrast, in the second example Alice has the information, the color Blue, in her subjective consciousness, which she then encodes into physical form using a pen and making symbolic markings onto paper. When Bob sees those physical markings his brain interprets them as the color Blue.

Notice that it is only when both Alice and Bob have the color blue in their subjective consciousness that it is meaningful. When that information is transferred to paper and takes physical form it is meaningless. It only becomes meaningful if we happen to know the particular coding system, ie the language.

To make the point more clearly consider these examples.

Четыре важных "нельзя" при эксплуатации или утечке газа в квартире:

Не ремонтируйте газовые приборы самостоятельно.

Не привязывайте к газовым трубам бельевых веревок и не используйте их в качестве заземления.

Не исправляйте сами дефекты газопроводных труб! Инструментом можно высечь роковую искру.

Не оставляйте без присмотра работающие газовые приборы, особенно - если доступ к ним имеют дети.

Hopefully, there will not be too many of you who read English, Russian and Japanese. So, to most readers the markings and symbols in at least one of these examples will be meaningless and yet, I assure you, they do both contain important safety information.

From all of this a curious fact emerges. When information takes physical form it is meaningless but when it ceases to have physical form, in our subjective consciousness, it becomes meaningful.

An implication from this is that everything we learnt at school and for a trade or profession was in the form of meaningless physical information. In the form that it became meaningful to us it ceased to physically exist. So, all of the information that we rely on to function does not physically exist.

4 The Seat of Consciousness

In order to solve the hard question of consciousness, ie why we experience consciousness in the way that we do, much of the research has been, and continues to be, focused on the painstaking work of unpicking the neural correlates of consciousness. That is to say, taking specific facets of consciousness and identifying the precise cluster of neurons in the brain responsible for producing that effect.

Consequently, it is the brain's cortex that is the focus of attention. This is hardly surprising since it has long been thought that consciousness is a peculiarly human attribute.

But might we be looking in the wrong place? The neurosurgeon Wilder Penfield conducted many operations on those suffering from epilepsy during which substantial volumes of the patient's cortex was removed whilst the patients remained conscious throughout.

Subsequently, together with his colleague Herbert Jasper, he proposed that "the highest integrative functions of the brain are not completed at the cortical level, but in an upper brainstem of central convergence supplying the key mechanisms of consciousness" (Penfield 1952 in Merker 2007).

This position might also be supported by those unfortunates born with hydranencephaly, where the two cerebral hemispheres are completely or largely absent and yet the patients present conscious awareness.

Further research, as documented by Merker (2007) has identified the superior colliculus located in the roof of the midbrain as playing the key integrative role in consciousness. Its physical structure, together with its connections to other regions of the brain equip it to perform three key behavioral decision making functions, motivation selection, target selection and action selection.

There can be conflicting motivators, eg fear, thirst, sex, and selection must be performed according to a priority system. Having determined the chosen motivator there may be multiple ways of satisfying it. For thirst, maybe there is a need to weigh the probability of finding a more plentiful and pleasurable drink at a greater distance against the certainty of finding a less plentiful and pleasurable drink nearby. And for action, maybe there are options for getting to the source of drink that need to be considered and a choice made.

Consequently, whilst hungry nematodes, lacking subjective experience, respond to starvation with a random rather than a directed search, hungry rodents, ants, and bees will navigate to places where they have previously encountered food. Their internal state of hunger triggers a highly directional food search, focused on locations where food was previously found.

Goal directed searching for food, as an indicator of subjective conscious experience, emerged before the evolutionary divergence of insects and vertebrates about 600 to 700 million years ago. All of this appears to suggest that subjective conscious experience emerged as a consequence of the transition from stationary plants to mobile insects.

It is also the most effective way of solving the reafference problem. When the eyes move the image of the external environment moves across the retina, which, unless corrected for, would mean that the external environment is moving.

Subjective consciousness corrects for this so that when we move our eyes we see that what we are observing does not move as a consequence. In effect, we create a spatial simulation as part of our subjective conscious experience.

Our spatial simulation is an illusion on a grand scale. It feels as though what we are experiencing is out there, outside of ourselves. But this cannot be true. To begin with there is significant processing involved for correcting for the reafference problem. This involves switching off the inputs from the optic nerve when moving our eyes or our heads, then filling in the blanks. A process called saccadic masking.

Then there is the matter of correcting for the time delay involved in processing the sensory inputs by creating a predictive simulation. If the result of all of this is outside of us, onto what is it projected? Clearly, we carry it around with us because it remains within our heads.

Consequently, that distant star we see at night is no further than the inside of our heads. The image of the star that we experience does not physically exist but is an optical illusion, rather like a hologram, that only appears to be distant.

5 Where do thoughts come from?

Earlier we learnt that the Ancient Greeks thought that information was created by events and that this became incorporated into Information Theory. But, perhaps that is not the whole story.

I chanced upon a web page containing the outline of a Philosophy text book being developed by Mark Mercer, Professor of Philosophy at St Mary's University, Halifax, Canada. Chapter 40, Mental Events and Physical Events, began with a simple exercise. The point of the exercise was to draw attention to the two streams of thought that occur concurrently in our minds.

A common experience for many of us is to drive to a routine destination whilst thinking about something other than driving. We maintain a constant visual awareness of our surroundings and respond appropriately to make turns as required, but at the same time we maintain a concurrent stream of thought on matters totally unrelated to driving.

We referred to the spatial simulation which is part of our subjective conscious experience and enables us to navigate to our destination. Incidentally, the reason we make appropriate turns without conscious thought is due to 'environmental cuing'. Basically, our non-conscious mind detects visual cues in our environment that prompt us to turn at the appropriate point in our journey.

So, we can easily account for the constant visual image, which, as we have seen, isn't actually constant but nonetheless appears so in our subjective consciousness. But what of the other, concurrent stream of thought? Where does that come from?

We have already established that when information is meaningful it doesn't physically exist. Where might we find a source of information that does not physically exist? To locate the potential source we need to delve into the most fundamental nature of physical reality.

Everything that we see, that physically exists, consists of atoms. The simplest atom is that of hydrogen, consisting of one proton in its nucleus and one orbiting electron. The proton contributes 99.95% of the mass of the atom. However, the proton, in turn, consists of three quarks held in place by gluons. The three quarks account for only 0.2% of the mass of the proton.

The remaining 99.8% of mass comes from the energy of the gluons, which themselves, paradoxically, have zero mass. Einstein showed that energy and mass have equivalence via his famous equation, $E = mc^2$. So, strange though it may seem, 99.3% of the mass of a hydrogen atom is contributed by particles with zero mass.

Also, gluons are virtual particles, so called because they are continually popping into and out of existence, existing for a mere 10^{-22} seconds. This begs the question where are virtual particles when they are not 'in existence'?

To answer this we turn to quantum physics and the Zero Point Energy Field. It is here that particles reside until they have sufficient energy to cross the threshold into our world. Strictly speaking they are not particles at all but rather vibrations. Since meaningful information does not physically exist and all that is in the Zero Point Energy Field does not physically exist, maybe it is populated exclusively by meaningful information?

There is no concept of space or time in the Zero Point Energy Field. So, bizarre though it may seem, in the Zero Point Energy Field the future has already happened. This is similar to the Block Universe concept, where all of space-time already exists.

Also, at the quantum scale, information can neither be created nor destroyed. Hence, all meaningful information exists from everywhere, including all of its past and all of its future, in the Zero Point Energy field.

The absence of space and time might account for apparent communication of information faster than the speed of light by entangled particles. It might also provide a basis for the concept of One found in some religions. In Hinduism and Buddhism, for example, everyone and everything is connected within the One.

Since all that physically exists in our world originates from the Zero Point Energy Field and our world has the emergent properties of space and time, there is no obstacle to information being accessible from a different place and at a different time, which might account for some paranormal phenomena, eg remote viewing and precognition.

An emergent property is one that is not fundamental, eg temperature. What we call temperature is actually a property that emerges from the velocity of atoms and molecules.

Sometimes, when we have been trying to solve a problem, inspiration comes when we are daydreaming, staring vacantly or performing a routine task. At such times the brain's Intrinsic Attention Network kicks in and knows the problem that we are trying to solve. In addition to the brain's own resources perhaps it can obtain inspiration from future information within the Zero Point Energy Field.

Given that all physical matter, including our brains, consists almost entirely of virtual particles popping in and out of the Zero Point Energy Field, might these same virtual particles be vectors of information between the Zero Point Energy Field and our brains?

Virtual particles are not actual particles but vibrations. As such, they have the potential to act like carrier waves, in much the same way as radio carrier waves with information encoded by frequency or amplitude modulation.

Perhaps a subcortical area of the brain serves the function, similar to a radio tuner, to extract the information from the virtual particle carrier wave.

I had been mulling over these issues whilst out walking. That was my concurrent thought stream. When I walked in my front door the name Tycho Brahe popped into my head. Why? Tycho Brahe was a 16th century Danish astronomer, noted for his meticulous astronomical observations. He also mentored Johannes Kepler, whose laws of planetary motion laid the foundation for Sir Isaac Newton's ground breaking discoveries. I hadn't been thinking of anything remotely connected to astronomy. It was a purely random thought. In the same way that the thought that triggered this investigation was purely random, ie "Blue does not exist".

Could it be that this meaningful information that does not physically exist originates from the Zero Point Energy Field, the repository of information that does not physically exist?

An alternative explanation for random thoughts is proposed by Erik Hoel, assistant professor at Tufts University. Artificial neural networks are trained on large data sets. However, the algorithms so derived would only apply to those specific data sets due to what is termed as 'overfitting'. Consequently, random data is introduced in order for the algorithm to apply to more general data. This might also account for the apparent random nature of dreams, according to his Overfitting Behavior Hypothesis (OBH).

"Once you adopt the notion that reality and information are the same, all quantum paradoxes and puzzles—like the measurement problem [...]—disappear."
Zeilinger quoted in Brockman, J.:What we believe but cannot prove: today's leading thinkers on science in the age of certainty (2006).

6 Subjective Consciousness and the Self

I had been given this question to consider, "if information and consciousness can each exist independently from a physical medium might they be related, and if so, how?"

Earlier we found that the seat of consciousness was located in the superior colliculus. That is to say that the superior colliculus is the specific brain region responsible for creating subjective conscious experience.

But that is only part of the story. Consciousness is multifaceted and we have only located two sets of functions, ie spatial simulation and behavioral decision making. Deductive reasoning, by contrast, has been found to occur in the frontal parietal lobes of the brain.

Also, memories are reconstructed when recalled, which can be a source for error, and become incorporated within subjective consciousness. There is also the matter of the concurrent thought stream. Where might that be located?

The concurrent thought stream occurs within our subjective conscious experience, which consists exclusively of meaningful information, ie information that does not physically exist. This leads to the inescapable conclusion that subjective consciousness itself does not physically exist. Might it also be somehow located within the Zero Point Energy Field?

Earlier we found that the spatial simulation that feels as though it is outside of us, in fact, remains within our heads. Paradoxically, the subjective conscious experience, which we intuitively feel is inside our heads might, in fact, reside outside of us in the Zero Point Energy Field.

Obviously, the 'outside' in which the spatial simulation appears to reside is different from the 'outside' where the Zero Point Energy Field is located. The former outside exists within our world whereas the latter doesn't. Also, it is important to distinguish between the physical brain region responsible for generating the spatial simulation and the non-physical experience of the spatial simulation itself.

Furthermore, earlier we found that all that exists in our physical world depends for its existence on virtual particles that originate from the Zero Point Energy Field, where they do not physically exist. So, everything stems from nothing.

This seems to say the same as the Big Bang Theory, part of the accepted Standard Cosmological Model. It is also, surprisingly, consistent with the Faith account for creation, ie that God created everything from nothing.

Since everything that physically exists consists mainly of virtual particles rapidly transferring between the Zero Point Energy Field and our world, this also applies to our physical bodies, and indeed our brains. Could it be that virtual particles are the carriers of information between the Zero Point Energy Field and our brains?

Our Self resides in the subjective consciousness. It's where we live our lives. Yet it is populated by that which does not physically exist. When you think about it, that which is most important to us, that has most significance in our lives, occurs exclusively within the subjective consciousness. So, all that is most important to us does not physically exist.

We have also seen that the subjective consciousness might, itself, be located in the Zero Point Energy Field. Since time does not exist in the Zero Point Energy Field this suggests that the Self might not be constrained to our mortal timespan. That is to say that the Self existed, but not physically, before we were born and continues to exist, but not physically, after we die.

I have repeatedly used the expression 'does not physically exist' rather than simply 'does not exist' for good reason. Clearly, our subjective consciousness and all that goes on in there does exist, just not physically.

Subjective consciousness, meaningful information and reality at its most fundamental, are all essentially abstract. So, alongside the physical domain is the non-physical abstract domain. The prevailing thinking is that the mind arises from the physical brain.

The existence of the abstract domain might only be conceded on the basis that it arises from the physical domain. However, it rather seems that the reverse might be nearer the truth and that it is the physical domain that arises from the abstract domain. In much the same way that our physical world emerges out of the Zero Point Energy Field, via virtual particles.

To quote Max Planck, regarded by many as the father of quantum physics, "I regard consciousness as fundamental. I regard matter as derivative from consciousness. We cannot get behind consciousness. Everything that we talk about, everything that we regard as existing, postulates consciousness."

That said, both the physical domain and the abstract domain benefit from each other. It is often the case that observations in the physical domain, when taken collectively, reveal an underlying abstract principle. The development of the laws governing planetary motion mentioned earlier are a good example of this. The abstract principle can then be applied to advantage in many physical situations, as Newton's Laws in Classical Physics continue to do so. Such is the power of abstraction.

As evidence of the physical domain arising from the abstract domain, consider the island of Manhattan. It was the abstract information in the subjective consciousness of city planners, architects, construction engineers etc, that built all of the very physical skyscrapers that stand tall on the island of Manhattan. What a testament to the power of Collective Consciousness!

I had known for some years that normal infant brain development required touch stimulation. On the face of it this might seem odd since about 83% of the information our brain receives comes via the sense of sight. Yet those born blind suffer significantly less cognitive impairment compared with those deprived of touch.

This might be due to the sense of sight evolving much later than that of touch. Touch is also important in establishing the link between the images in our subjective consciousness that do not physically exist and the real world objects that do physically exist. Also, by having to reach out or maybe crawl towards an object before receiving its touch stimulation babies and infants learn the concepts of both space and time.

Because we experience subjective consciousness, including spatial simulation, in the way that we do we might think that the sense of sight is a prerequisite. This cannot be true when we consider other animals.

There is a saying, "blind as a bat", referring to the fact that bats lack the sense of sight. Yet bats perform a daily routine of leaving their roost around dusk in search for food then later successfully navigating their way back to their roost. This requires spatial simulation, which suggests that subjective consciousness might have been possible before the sense of sight evolved.

An ant's brain is smaller than the size of a pinhead yet it too experiences spatial simulation. Impressively, it uses a combination of the Sun and memories of landmarks to steer a bearing, irrespective of its own body orientation, back to its nest.

7 Who am I?

I remember reading in a novel the expression, "Conscious times three", meaning that the individual could correctly answer three simple questions, "Who am I? Where am I?" And, "When am I?" (What day, date and time is it?)

"Know thyself."

(Inscription at temple of Delphi 4th century BCE)

"To thine own self be true."

(Polonius, Act 1 Scene 3, Hamlet, William Shakespeare)

When a pilot flies in cloud he uses his instruments to infer how his aircraft is behaving and, therefore, whether any corrective inputs are required to maintain the desired heading, altitude and airspeed.

It seems that we humans are engaged in a similar but opposite procedure. We use our behavior to infer what that tells us about the Self that is in control. Let me explain.

We make conscious decisions about major events, eg moving house, changing jobs, using what Daniel Kahneman referred to as the slow decision-making system. But, as he discovered, we use the fast, intuitive, non-conscious decision-making process to orchestrate our actions, our behavior. He also noted that we might use the slow system to try to account for the decisions made by the fast system, but we cannot know how the fast system arrived at its decisions, and so we confabulate an explanation.

Since it is our non-conscious mind that is orchestrating our behavior it is our non-conscious mind that is in control. But, as we have seen, we cannot know how it makes its decisions. We cannot know what values, rules or policies it is using. In short, we cannot directly know anything about who's actually in control of our lives. Consequently, any notions we might have about our true Self can only be inferred by observing our behavior.

Perhaps there isn't a single Self, but two quite separate Selves.

The Self that we think of as being our Self resides in our conscious awareness. Our conscious awareness is where we experience the world outside and conduct our abstract, rational reasoning. But let's look closer. These two functions are referred to as the Extrinsic Attention Network and the Executive Control Network respectively. Both functions are conducted in the frontal parietal lobes of the brain, and in so doing compete for resources. This is evidenced by the classic case of an individual having to stop walking in order to count down from 100 in 7's.

What we experience as the conscious Self is when we are observing the world outside, ie when the Extrinsic Attention Network is engaged. To the extent that the Executive Control Network is engaged the observing Self is disengaged. So, the conscious Self is essentially a Passive Self.

On the other hand, our actions are orchestrated by the Self that resides within the non-conscious mind. We cannot directly know anything about this Self. We can only infer from observing our actions what sort of person this

Self actually is. Put bluntly, it is impossible for us to answer the simple question,

"Who am I?"

Because it drives our actions, this is the Executive Self. Consequently, it could be argued that our true Self is the unknowable Executive Self, whereas the Self we think we know is merely an observing, Passive Self.

Also, the Passive Self is almost certainly not the same as the Executive Self. Daniel Kahneman found that financial traders did not behave in the way that they thought they did. Similarly, Dennis Shaffer found that baseball catchers didn't behave in the way that they thought they did.

This tells us that, in both cases, the underlying values dictating their behaviors were not the same as they thought they were. In other words, the Executive Self was not the same as the Passive Self.

This view is consistent with the model of reincarnation, where our real life is the life lived between mortal lives and where the life goals for each incarnation are set. These goals are downloaded with the Self into the mortal body and so are known to the Executive Self but hidden from the Passive Self.

The following chart shows the relationship between the various entities in both the material and the immaterial domains.

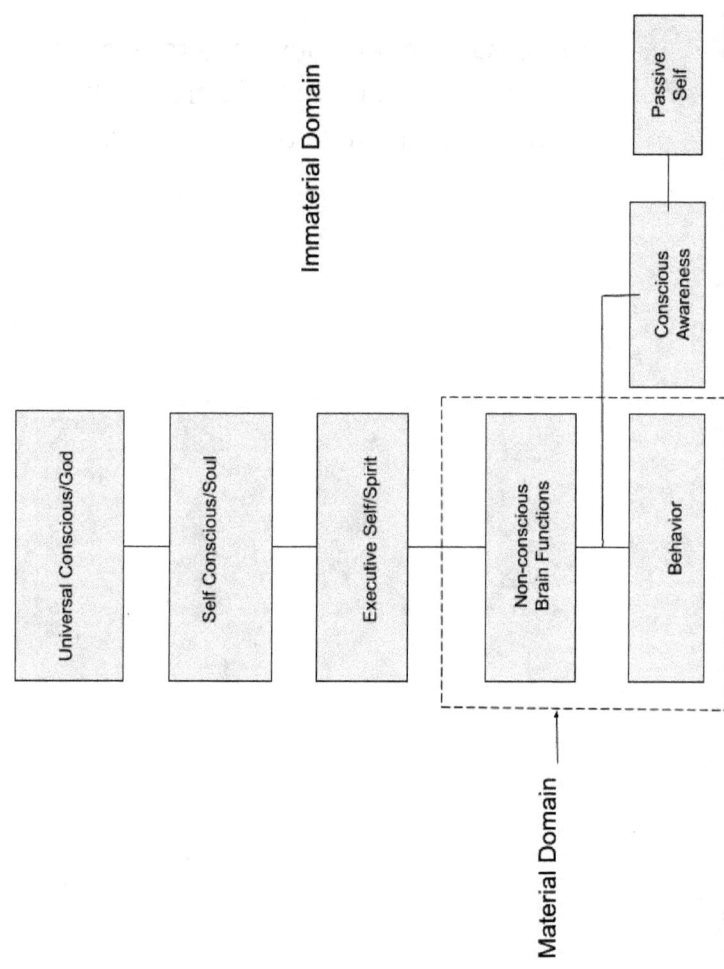

Immaterial Domain

Universal Conscious/God

Self Conscious/Soul

Executive Self/Spirit

Non-conscious Brain Functions

Behavior

Conscious Awareness

Passive Self

Material Domain

48

Free Will?

I had been aware for some time that decisions were made by our non-conscious mind and that we subsequently felt that we had made a conscious decision. Consequently, it seemed that free will could not be exercised by the conscious mind, but rather the non-conscious mind.

However, the fact that our conscious mind does not exercise free will does not necessarily mean that our non-conscious mind does make free will choices. Let me explain.

Free Will and Determinism are mutually exclusive.

Free Will - the power of acting without the constraint of necessity or fate; the ability to act at one's own discretion.

Determinism - the doctrine that all events, including human action, are ultimately determined by causes regarded as external to the will.

Experiments using fMRI (functional magnetic resonance imaging) have shown that it is possible to predict a motor decision, eg moving an arm, up to eleven seconds before the conscious decision to do so, and up to four seconds before an abstract decision. This suggests that brain function, including decision making, is fundamentally deterministic. Consequently, the brain cannot initiate a free will choice, itself.

It seems reasonable to suppose that all physical processes are governed by the laws of cause and effect, and so are fundamentally deterministic. The brain is, indubitably, a physical organ, so it is not surprising that its function is deterministic.

However, as we have seen, free will could be exercised in the immaterial domain, eg by the Executive Self / Spirit. A decision here might initiate physical non-conscious brain function, which the conscious mind subsequently becomes aware of and 'claims' as its own decision.

If the immaterial domain exercises free will, because free will and determinism are mutually exclusive, it follows that the immaterial domain cannot involve determinism, ie no chain of cause and effect. Because we live in a physical world, bound by the laws of cause and effect, it is difficult to comprehend an environment that is totally based on free choice.

Another way to view it is to consider the non-physical domain driven by will alone. This view resonates with the Faith perspective that everything was created and is sustained by the will of God.

8 Where is Memory?

Where is memory stored?

"These data are in accordance with the idea that the posterior hippocampus stores a spatial representation of the environment and can expand regionally to accommodate elaboration of this representation in people with a high dependence on navigational skills." (Maguire and Gadian, 2000)

"Higher encoding-state estimates from stimulation were associated with greater evidence of neural activity linked to contextual memory encoding. In identifying the conditions under which stimulation modulates memory, the data suggest strategies for therapeutically treating memory dysfunction." (Ezzyat and Kragel, 2017)

Both of the above examples seem to suggest that memory is stored in the brain. In the case of spatial memory, specifically within the posterior hippocampus.

However, the following seem to contradict this position, since memory appears to have been stored without a brain, a physical body or anything physical at all.

Life Before Life: Childhood Memories of Previous Lives by Jim B Tucker, and Reliving Past Lives: The Evidence Under Hypnosis by Helen Wambach.

Also, Martin Scorsese's 'Kundun'. A biopic in which artefacts belonging to the deceased Dalai Lama, mixed with other artefacts that didn't belong to the deceased Dalai Lama, are laid out in front of a four-year old child who unerringly selects only those belonging to the deceased Dalai Lama by pointing and saying, "Mine!".

How might these two positions be reconciled?

Obviously, there might be two completely separate and independent ways of storing memory. However, there might be an alternative explanation; one based on our everyday experience. We have already established that all of our experiences, including recalled memories, though they may be triggered by an external physical entity, do not, themselves, physically exist.

Whenever information is communicated between two fundamentally different types of medium a conversion process has to happen. In this instance we are talking about communication between a physical brain and a non-physical domain in which experience, including recalled memories, occurs.

We experience something similar when we use a Self-Checkout in the supermarket or a Cash Dispenser. We use a touch-sensitive screen to indicate our choices. The computer encodes a range of symbols, which the screen displays and our brains interpret to give meaning to them. We decide which one we want and indicate our choice by touching the relevant symbol.

The generic name for this interaction is via a GUI, a graphical user interface. By this means the meaningless physical information within the computer is communicated to the meaningless physical information in our brains. However, the meaningless physical information in our brains has to undergo some form of conversion to become the meaningful information in our memories.

This suggests that the meaningful information is not actually stored in our physical brains but in the immaterial Quantum Zero Point Energy Field. So, the brain, itself, instead of actually storing memory, acts as a converter to convert the meaningless physical information encoded in our brains into the meaningful information that our conscious mind uses. We could say that the brain acts as a QUI, a quantum user interface.

The experiments that appeared to indicate that memories were stored in the brain established the site of encoding and assumed physical storage, but the encoding could equally have been like the computer's encoding of a symbol in a GUI. In the case of the brain the encoding acts as the interface to the quantum domain.

9 What is imagination?

There are several ways to answer this question. To begin with we could look at a definition, but let's look at a definition of what it is not first.

Consciousness

the state of being aware of and responsive to one's surroundings.

With consciousness, it's all about sensing and responding to our environment. Imagination is definitely not that.

Imagination

the faculty or action of forming new ideas, or images or concepts of external objects not present to the senses.

Notice how this definition is still anchored on external objects, it's just that they're not present in our environment at that moment in time. But we know that imagination is so much more than that. Imagination can include objects that have never existed in our environment. Frank Whittle imagined a jet engine that, up until then like all inventions, had not existed in anyone's external environment. And it doesn't have to be an object. Composers imagine music that has never been heard in anyone's external environment.

Or, indeed, anything that can be sensed at all. Einstein imagined a concept of space and time in a completely original way that resulted in his laws of Relativity.

So, an important way of thinking about what imagination is is to consider the potential content of imagined experiences. But we need to go deeper and ask ourselves, "of what substance is imagination?" If, indeed, imagination has any substance at all.

We quickly hit the same problem that we had at the outset with the color Blue. Though we intuitively feel that what we imagine is in our brains, if we were to look at our brains whilst we were imagining we would obviously not see the content of our imagination.

However, using fMRI we could observe the brain activity associated with imagining.

Indeed this has been used to enable those diagnosed with persistent vegetative state to give 'yes / no' responses to questions, eg by imagining playing tennis.

Whilst this shows the brain processes involved in imagination it gives no clue to the content, except by association. This is because, like the color blue, the content itself does not physically exist. What the brain is doing is acting as the interface between the physical encoding of the information and the non-physical information in the experience of imagination.

10 The Hard Problem

Whilst neuroscientists progress the field of neural correlates of consciousness, ie the group of neurons involved in a specific aspect of consciousness, the hard problem of consciousness stubbornly remains. How is it that our non-physical subjective experience of consciousness is created from our material brain?

The prevailing scientific view is that the mind is a construct of the brain, in keeping with a generally monism worldview. By contrast dualism, after René Descartes, takes the view that the mind is non-physical and separate from the brain.

Consequently, I had thought that monism and dualism were mutually exclusive, rather like determinism and free will. However, on further examination, this might not be the case.

Monism - a theory or doctrine that denies the existence of a distinction or duality in a particular sphere, such as that between matter and mind, or God and the world.

Dualism - denotes either the view that mental phenomena are non-physical, or that the mind and body are distinct and separable.

Whilst, on the face of it, these definitions imply that they are indeed mutually exclusive, one interpretation of monism, neoplatonism, is that everything is derived from the One. This is a view shared both by Eastern Faiths and by Science, in the Big Bang theory.

The neuroscientist, Anil Seth, an advocate of the mind as a construct of the brain, and hence a monist, describes consciousness as "controlled hallucination". The hallucination, itself, has no material substance, which seems to imply dualism.

Whilst I subscribe to the concept of the One, the evidence of our own experience suggests that the mind and consciousness are non-physical. Hence, I find myself in the uncomfortable position of being a dualist within monism.

I do not totally deny that the mind is a construct of the brain, but that is only one part of the story. Regarding the material world around us, our brains construct our non-physical experience of it. Conversely, thoughts emanating from our non-physical consciousness take on physical form in the neurons in our brains, and might, thereby, affect the external material world.

This transposition, or interface, between the material brain and the immaterial consciousness lies at the heart of the hard problem of consciousness.

Our material brains are, ultimately, composed of virtual particles, and these same virtual particles are, in effect, vibrations transiting between the physical world and the non-physical Quantum Zero Point Energy Field, where consciousness, itself, might reside.

Modulation of these vibrations might carry information about the material world from our brains into the immaterial consciousness domain, and information about immaterial thoughts from our non-physical consciousness into our physical brains.

So, the hard problem of consciousness might be solved by experimentally determining whether modulation of virtual particles / vibrations in the brain occurs, thereby carrying information in either analogue or digital form between the physical domain and the non-physical domain.

Earlier we saw that the key behavioral decisions of motivation selection, target selection and action selection were performed by the subcortical superior colliculus.

Consequently, it is tempting to think that this is the location of the interface 'tuner' between the physical domain of the brain and the non-physical domain of conscious experience. Rather like the resonant quantum tuner believed to be at the heart of avian navigation.

However, it seems more likely that motivation selection is performed in a deterministic manner based on interoception, based on the body's own needs at that instant in time. That still leaves the possibility that target selection and action selection might be open to free will choices made in the non-physical domain.

11 Consciousness Creates Time

I was pondering the question, "In what context does the material become the immaterial other than consciousness and virtual particles?" This was inspired by recent research into the fundamental mechanisms for the formation of memory.

Initially, I had thought that it undermined the concept of immaterial experience but, of course, it doesn't. It is still only concerned with the physical representation of memory, not the experience of memory.

So, the long-standing puzzle remains, how does the material neural correlate of memory give rise to the immaterial experience of memory. By searching for other contexts in which this transition from material to immaterial occurs I was hoping to throw some light onto the problem.

However, as Max Planck said, "I regard matter as derived from consciousness." So, let's start from this position. When I investigated the statement "Blue does not exist", I found that it was true, in the sense that it does not physically exist, by tracing photons all the way from their source in the sun to the stimulation of the ventral occipital lobe in the brain where color is determined. Nowhere in that journey does the color blue physically exist. The experience of the color blue is, in a sense, an optical illusion, an immaterial experience.

Taking Max Planck's statement as the starting point means beginning with the immaterial experience of the color blue and tracing it back via its material source in the ventral occipital lobe and then all the way back to the sun.

Generalising from this we have all of our collective consciousness causing the creation of sequences of events in reverse order. This gives the impression of a forward arrow of time creating an immaterial experience from a material origin. Instead, it is the immaterial experience creating a backward chain of events such that the required effect creates the necessary cause.

So, instead of

Sun -> Sky -> Eye -> Brain -> Experience Blue

We have

Sun <- Sky <- Eye <- Brain <- Experience Blue

This means that the fundamental reality is, in effect, a backward arrow of time where cause is created so as to achieve the required effect.

The sequence is initiated when the last cause created is activated. This produces the familiar forward arrow of time with cause preceding effect. Consequently, we experience the illusion that it is the material creating the immaterial, when in fact the reverse is true and it is the immaterial that is creating the material, and as a byproduct consciousness creates time.

This paradigm of effect preceding cause is described as retro-causality and has been proven experimentally specifically with regard to non-conscious decision making (choosing) of non-conscious future (negative or neutral) stimuli. This was reported in the paper Time and Consciousness (Maier and Buechner 2016), which also concludes that consciousness creates time.

In addition, it was the apparent acausal random time interval in radioactive decay that inspired

Yakir Aharanov to see effect preceding cause as re-establishing a causal relationship, and resulted in time-symmetric quantum mechanics.

Of course, if we follow this line of argument to its logical conclusion it would mean that the very beginning of time ultimately came about in order to achieve the very end of time. That the end of time and the beginning of time are causally connected is not an original idea.

"I have made known the end from the beginning."

(Isaiah 46:10)

Interestingly, the Hebrew word translated here as 'made known' is translated elsewhere as fashioned or accomplished. So, this text could equally read 'I have fashioned the end from the beginning', which suggests a form of Intelligent Design that embraces evolution.

12 Purpose vs Effect

I had attended an online talk given by Nobel Laureate Sir Paul Nurse, which I found both enjoyable and informative. However, in the subsequent Q&A session he referred to purposeful behavior of living entities. I have a problem with the use of purpose in this context, or possibly any context. Let me explain by the following scenario.

Feeling hungry, and as it was approaching lunchtime, I decided to make myself a sandwich. That was when I noticed that I was getting low on bread, so I made a mental note to go and buy a loaf from my local supermarket. Later that day, as planned, I set off to the local supermarket. Had someone stopped me to ask, "why are you going to the local supermarket?" I would have replied, "to buy a loaf of bread". From this they would have inferred that the purpose of my going to the local supermarket was to buy a loaf of bread.

But that isn't what actually happened. What actually happened was that hunger caused me to make myself a sandwich. Making myself a sandwich caused me to notice that I was getting low on bread. Getting low on bread caused me to go to the local supermarket. It was a chain of causation that caused me to go to the local supermarket. Purpose did not feature anywhere in the sequence of what actually happened. Purpose was inferred after the event. This leads me to suggest that there is no such thing as purpose, only cause and effect.

That said, during my career I designed computer systems. The very first statement in the documentation was entitled 'Purpose'. This was so important since it guided everything that the system set out to achieve and was the fundamental measure against which success might be judged. Another word for Purpose in this context might be 'Effect'. In other words, what effect should this system have on the business? We are all familiar with Effect due to its relationship with Cause. We are used to cause preceding effect.

However, in the case of systems design the effect we wish to achieve determines the preceding chain of causation. This is similar to the fundamental reality of consciousness creating the chain of causation necessary to arrive at the desired effect and is consistent with, "I have fashioned the end from the beginning". (Isaiah 46:10)

13 Why is Blue Blue?

Neuroscientists will say that we experience the color blue because of the way our bodies and brains process the photons of light entering our eyes, vibrating at that particular frequency (6.66×10^{14} Hz).

But I was reminded of the account of Ian McCormack who suffered five box jellyfish stings whilst diving at night off the coast of Mauritius. The venom from the box jellyfish is so powerful that just one sting can be sufficient to cause death within minutes.

He died in hospital whilst being treated with anti-toxins and was dead for fifteen minutes. During this time, without the use of either eyes or brain, he 'saw' green grass and a blue sky. So it seems that the sky is perceived as blue whether with or without the use of our eyes and brain.

Of course, some will discount his account as fiction or imagination, in which case it tells us nothing. However, if true, it surely must tell us something, but what exactly?

Ian McCormack's experience was in a domain after life, ie a non-physical domain. It is interesting that both Faith and Science have a role for a non-physical domain. For Faith it is a spiritual domain where, for many, God resides. In Science it is the Quantum Zero Point Energy Field from which every physical thing is continuously created. Might Faith and Science be like the tale of the blind men and the elephant, effectively viewing the same domain from completely different perspectives?

In any event, it is clear that the laws of Physics, as we understand them, no longer apply within the non-physical domain. So, we have two totally different domains within which the sky is consistently blue. But is the sky blue in the non-physical domain because Physics makes it so in the physical domain or vice versa? In other words, which came first?

In the non-physical domain the law of cause and effect no longer applies. So there can be no determinism there. Consequently, it is a domain dominated by Will. (Free will and determinism are mutually exclusive by definition). So, the sky is blue in the non-physical domain because it is willed to be so.

As we have seen, the physical domain is continuously created from the non-physical domain. So, the sky is blue in the physical domain because it is blue in the non-physical domain. And the sky is blue in the non-physical domain because it is willed to be so. Hence, we can say that the sky is blue in the physical domain, ultimately, because it is willed to be so. Therefore, Blue is Blue because it is willed to be so.

14 Blue light is important

Although the color blue does not physically exist, nonetheless blue light plays an important role in nature.

Lauren Foley, University of Massachusetts Medical School, works with fruit flies (Drosophila) that normally sense magnetic fields using cryptochrome. Cryptochromes are proteins found across biology in insects, birds and mammals, including humans.

They have a wide range of functions, including regulating plant growth rates, biological clocks and enabling avian navigation. They are thought to sense weak magnetic fields in many species, through a quantum mechanism in which the Earth's magnetic field alters the rate at which the protein is activated by light.

She proved that fruit flies navigate by magnetosensing by placing them in an artificial magnetic field and training them to fly in a specific direction to search for food. Normal fruit flies can do this easily. Modified fruit flies that don't have the CRY gene, which makes the cryptochrome protein, cannot find their way to their meal.

When she loaded her modified fruit flies with the human CRY2 gene, she found that they were able to sense magnetic fields like normal fruit flies and so found their meal. However, the flies' magnetic sense was only restored when they were bathed in blue light.

Margaret Ahmad at Sorbonne University in Paris, France, and her colleagues exposed thale cress seedlings (Arabidopsis thaliana) to weak pulses of radio frequency (RF) radiation at 7 megahertz.

Ahmad, who discovered cryptochromes in the 1980s, wondered if these receptors might also be sensitive to radio waves. Extremely weak RF radiation is known to disrupt magnetosensing in birds, insects and rodents, but the mechanism is unknown.

The team predicted that if the quantum cryptochrome theory is correct, RF radiation should also interfere with the sensor, blocking the effect of Earth's magnetic field. This is indeed what they found, with the seedlings responding in the same way as a control group placed in a null magnetic field.

So we have cryptochrome sensitive to blue light (610 to 670THz) causing the quantum photoelectric effect of electron displacement. Thereby making it sensitive to Earth's magnetic field. But separately sensitive to RF at 7 MHz causing the cryptochrome to be insensitive to Earth's magnetic field. So, RF at 7 MHz inhibits the quantum photoelectric effect in cryptochrome.

For humans, blue light has also been associated with increased alertness. When a bar was designed with subdued blue lighting it had the opposite effect to that expected. Instead of creating a subdued atmosphere customers became more alert. Research evidence supports this effect.

15 Conclusion

Having begun with the simple statement, "Blue does not exist", we have found that information is only meaningful when it does not physically exist. Furthermore, we have seen that everything that we experience does not physically exist, and that it is only by the association of touch as babies and infants that we assume thereafter that our experiences are about real physical objects.

Also, the Self, or what those of faith call the Soul, living as it does within the subjective consciousness, which in turn dwells within the Quantum Zero Point Energy Field, without constraints of time or space, exists before, during and after mortal life.

In addition, all that physically exists is ultimately dependent for its existence on virtual particles, popping into and out of existence from within the Quantum Zero Point Energy Field. This is not only consistent with the Big

Bang Theory but also the faith account of everything being created from nothing.

The Zero Point Energy Field might also be known as the Universal Consciousness. Hence, the Universal Consciousness might be the source of all physical matter, just as Max Planck predicted.

It might also explain why consciousness appears to interact with matter at the quantum level, according to the von Neumann and Wigner explanation of Niels Bohr's observer effect in the double slit experiment.

This, or a similar, mechanism might also be how free will choices made by the Executive Self / Spirit initiate non-conscious brain function, leading ultimately to the experience of a conscious decision.

Finally, in order to account for the apparent creation of the immaterial from the material, we have the immaterial creating the material, fixing the color blue as blue and by a reverse sequence of events, as a byproduct, consciousness creating time itself.

Bibliography

Ardiel, E. L. and Rankin, C. H. (2010) The importance of touch in development, Paediatric Child Health, 2010 Mar;15(3):153-156.

Battersby, S. (2008) It's confirmed: Matter is merely vacuum fluctuations, New Scientist 20 November 2008.

Bekenstein, J.: Information in the Holographic Universe. Sci. Am. 289(2), 58–65 (2003) p63. https://www.scientificamerican.com/article/infor mation-in-the-holographic-univ/

Blackmore, S. (2002) The Grand Illusion: Why consciousness exists only when you look for it, New Scientist, 22 June 2002, p 26-29.

Briggs, H. (2017) Ants use Sun and memories to navigate, BBC News, Science and Environment, 19 January 2017. https://www.bbc.co.uk/news/science-environme nt-38665058

Collini, E., Wong, C., Wilk, K. *et al.* (2010) Coherently wired light-harvesting in photosynthetic marine algae at ambient temperature. *Nature* 463, 644–647 (2010).

https://www.nature.com/articles/nature08811

Davies, P., Gregersen, N.H. (eds.): Information and the Nature of Reality: From Physics to Metaphysics, Canto Classics edn. Cambridge University Press, Cambridge (2014)

De Graaf, T. A. and Sack, A. T. (2014) Using brain stimulation to disentangle neural correlates of conscious vision, Frontiers in Psychology 23 September 2014.

Ezzyat Y, and Kragel J E, et al (2017) Direct Brain Stimulation Modulates Encoding States and Memory Performance in Humans, Current Biology 2017 May 8;27(9):1251-1258. doi: 10.1016/j.cub.2017.03.028. Epub 2017 Apr 20.

Ghose T (2015) Birdbrains? Hardly: Baby Chicks Know How to Count, Live Science 29 January 2015.
https://www.livescience.com/49633-chicks-count-like-humans.html

Glattfelder, J. B. (2019) Information–Consciousness–Reality How a New Understanding of the Universe Can Help

Answer Age-Old Questions of Existence, Springer Open, the Frontiers collection.

Goldhill, O. (2019) Neuroscientists can read brain activity to predict decisions 11 seconds before people act. Quartz online March 9, 2019.

file:///media/fuse/drivefs-a20ef0b59520ee1f3ef3 1b0facb466b2/root/Consciousness/Neuroscien tists%20read%20unconscious%20brain%20act ivity%20to%20predict%20decisions%20%E2% 80%94%20Quartz.mhtml

Hoel, E. (2020) Dreams: The fictions we conjure while we sleep may do something far more powerful than reinforcing learning. New Scientist, 7 November 2020, p34-38.

Jaeger, G.: Entanglement, Information, and the Interpretation of Quantum Mechanics. Springer Science & Business Media, Heidelberg (2009) p234f

Kanai, R., Chang, A. et al (2019) Information generation as a functional basis of consciousness, *Neuroscience of Consciousness*, Volume 2019, Issue 1, 2019, niz016, 29 November 2019

Koenig-Robert, R. and Pearson, J. (2019) Decoding the contents and strength of imagery before volitional engagement. Nature, 5 March 2019, *Sci Rep* 9, 3504 (2019). https://doi.org/10.1038/s41598-019-39813-y

Kupers, R., Beaulieu-Lefebvre, M. et al (2011) Neural correlates of olfactory processing in congenital blindness, Neuropsychologia 49 (2011) 2037–2044.

Lau, H. C., Rogers, R. D., et al (2007) Manipulating the Experienced Onset of Intention after Action Execution, 2007 Massachusetts Institute of Technology, Journal of Cognitive Neuroscience 19:1, pp. 1–10

Maguire, E. A.,and Gadian, D. G., et al (2000) Navigation-related structural change in the hippocampi of taxi drivers, PNAS April 11, 2000 97 (8) 4398-4403.

https://doi.org/10.1073/pnas.070039597

Marcer, P.J. (1998). A Quantum Mechanical Model of the Evolution of Consciousness

Marchant, J. (2020) Plant protein responds to radio waves by making seedlings grow faster. New Scientist 14 August 2020.

Marchetti, G. (2018) Consciousness: a unique way of processing information. Cognitive Processing 19, 435–464 (2018). https://link.springer.com/article/10.1007/s10339-018-0855-8

Massie, E. W. (2012) Gravity and Zero Point Energy, Physics Procedia 38 (2012) 280-287.

McCormack, I. (2014) The 1988 Original Testimony. https://www.youtube.com/watch?v=Yg4188Ii-fg

McTaggart, L.(2002) The Field: The Quest for the Secret Force of the Universe, HarperCollins, London.

Maier M. A. and Buechner V. L. (2016) Time and Consciousness, in Nadin M. (ed) Anticipation Across Disciplines, Cognitive Systems Monographs 29, Springer International Publishing, Switzerland 2016.

Merker, B. (2007) Consciousness without a cerebral cortex: A challenge for neuroscience and medicine, The Behavioral and brain sciences, 2007 Feb;30(1):63-81; discussion 81-134.

Mitchell. E. D. and Staretz R. (2011) The Quantum Hologram and the Nature of Consciousness, Journal of Cosmology, 2011, Vol. 14.

Oldershaw, R.L.(2009) Towards a Resolution of the Vacuum Energy Density Crisis, arxiv.org/pdf https://arxiv.org/pdf/0901.3381.pdf appeals to the application of self-similar cosmological paradigm.

Penfield, W. & Jasper, H.H. (1954). Epilepsy and the functional anatomy of the human brain. Boston: Little, Brown & Co.

Peperrell, R. (2018) Consciousness as a Physical Process Caused by the Organization of Energy in the Brain, frontiers in Psychology, 1 November 2018.

Phipps-Nelson, J. and Redman, J. R. et al (2009) Blue Light Exposure Reduces Objective Measures of Sleepiness during Prolonged Nighttime Performance Testing. Chronobiology International, Vol 26 2009 - Issue 5.

Rees, G. (2007) Neural correlates of the contents of visual awareness in humans, Philosophical Transactions B Royal Society

London Biological Sciences, 2007 May 29; 362(1481): 877–886.

Siegel, E. (2016) Where does the mass of a proton come from? Forbes, August 3 2016.

https://www.forbes.com/sites/startswithabang/2016/08/03/where-does-the-mass-of-a-proton-come-from/#3d3f47c72e1d

Siva, N. (2006) Patient in Coma Plays Tennis: Researchers have shown that a patient diagnosed as being in a vegetative state does have conscious thought. Lancet Vol 5, Iss 11, P906, November 01, 2006. https://www.thelancet.com/journals/laneur/article/PIIS1474-4422(06)70592-0/fulltext

Stark, C. E. L. Okado, Y. and Loftus, E.S.(2010) Imaging the reconstruction of true and false memories using sensory reactivation and the misinformation paradigms, Learning Memory 2010 17:485-488.

Tucker, J. B. (2005) Life Before Life: Childhood Memories of Previous Lives, St Martin's Griffin, New York.

Vidral, V. (2010) Decoding Reality: the universe as quantum information, Oxford University Press, Oxford, UK.

Wambach, H. (1978) Reliving Past Lives: The Evidence Under Hypnosis, Harper & Row, New York

Wheeler, J.A.: Information, physics, quantum: the search for links. In: Zurek,W.H. (ed.) Complexity, Entropy, and the Physics of Information. Westview Press, Boca Raton (1990)

Yong, E. (2011) Humans have a magnetic sensor in our eyes but can we detect magnetic fields? Discover Magazine June 21 2011.

Zyga, L. (2011) Quantum no-hiding theorem experimentally confirmed for first time, Phys Org, March 7, 2011. https://phys.org/news/2011-03-quantum-no-hiding-theorem-experimentally.html